Bad Apples, Bad Water

Lead Arsenate Pesticide, the Holocaust, and Beyond

By Chris Baba

First Edition

Townies With Torches Press

This book is published by Townies With Torches Press at 489 Hobnail Ct., Frederick MD 21703. The author can also be contacted by email at cbaba001@comcast.net

Townies With Torches Press™

Paperback Version

ISBN

979-8-9945662-2-0

Library of Congress control number (LCCN) 2026901308.

Dedication

This book is dedicated to my Mom and Dad, wife, and all the soldiers and citizens of Ukraine who are fighting for their freedom against the tyranny of a ruthless Russian dictator.

Preface – How I became interested in lead.

As a biochemist I had a thirst for knowledge and a desire to figure out how things work. My interest in lead exposure had nothing to do with anything I was working on in my professional work, but came from the pages of a vintage book about airplanes written in 1925.[1]

This book described the great advances in the development of airplanes in the early 1920's. There was one page about the new technique of crop dusting and aerial spraying. There were descriptions of aerial crop dusting using lead arsenate, and other arsenic containing compounds, on apple trees, orange groves, and cotton fields. One public demonstration in Georgia drew a crowd of ten thousand people.

At first I was completely stunned by the fact that spraying heavy metal compounds containing lead and arsenic on food crops was considered normal. In fact, the book made it sound like this was a new miracle technology that would vastly increase crop production by reducing crop losses due to pests like the vegetable weevil and boll weevil.

I did some research on the internet and discovered that lead arsenate was used world-wide in the first half of the 1900's. However, it took a lot more research to discover that the amounts of lead arsenate pesticides used was enormous. The large amounts used can easily have lasting effects for centuries, and are probably contaminating the aquifers that feed well-water supplies today.

I knew that lead exposure had been studied extensively due to the sale of leaded gasoline, which is now almost completely banned by every country in the world. There are many studies that have linked lead exposure to a decrease in intelligence and an increased propensity to violence. This brought me to the hypothesis that lead exposure is probably one of the leading causes of war and violence.

--Chris Baba, December 25, 2025
  Frederick, Maryland USA

Table of Contents

Chris Baba

Chapter 1 - Overview

It is often said that hindsight is 20/20, but hindsight is blind in the darkest corners of the past unless we shine some light on it. This book will attempt to shine some light on one of the darkest corners of the past the world has ever known. If this were a book of fiction it would not sell well because it would be too hard to believe that the fate of a hundred million people rested in the hands of two scientists and one decision. Yet this is a true story. These events really happened in the early 1900's. And it is a decision that will live in infamy.

It was a time of great hope now that The Great War (WWI) was over. New advances in science, electricity, aviation, and manufacturing were fueling optimism about the future. No one wanted to reminisce about the carnage of the war when there were so many promising advances in science, farming, and manufacturing to work on.

Two scientists at the USDA (US Department of Agriculture) were testing the effects of a pesticide on orange trees. There was hope that it could increase the yield of oranges and orange juice without decreasing the amount of vitamin C in the fruit. Fruits containing vitamin C were urgently needed to support the nutritional needs of the rapidly growing population.

The great hope and optimism of the times probably influenced these scientists as they made their fateful decision. It was definitely the wrong decision, yet we are only now starting to realize that the correct decision probably would have prevented the Second World War, the Holocaust, the Spanish Civil War and the nuclear bombs that ended WWII. A hundred million lives might have been saved.

Back in the lab, these two scientists (E.M. Nelson and H.H. Mottern) had a dilemma. The oranges from trees that were heavily sprayed with the pesticide were being compared to the oranges from untreated trees. Their previous results showed no cause for concern, so they were surprised by the tests results from their animal testing. Two of the 12 guinea pigs that were fed orange juice from pesticide treated trees had died, while the other guinea pigs were fine.

If they were honest they would have published these results and raised the alarm about this arsenic based pesticide (lead arsenate) being poisonous and that it wasn't being sufficiently removed by the water bath washing technique that was recommended by the USDA at the time. No one knows the reason why they didn't do that. It is one of those dark corners of history that we can't shine a light on and analyze it with the 20/20 vision of hindsight. However, we know that they made the infamous decision to obscure the real results because they published it, but they didn't know that it would fatally change the fate of 100 million people.

Instead of comparing the treated animal group (with 2 dead animals) to the untreated group (with no dead animals) they decided to make a new fake "control" group that were fed a diet that was lacking any vitamin C. All of these animals died. They discarded the data from the 2 dead treated animals and implied that the treated oranges were safe because these animals fared better than the animals in the fake "control" group, who all died.

In hindsight this decision had consequences far greater than anyone could imagine.

These two scientists from the USDA were not the only ones who covered up the dangers of lead arsenate pesticide, and they died before the full consequences of their infamous decision would be realized. However, this coverup still continues today by some politicians, farmers and landowners who downplay the dangers of the poison that was applied to their lands.

This book is not fiction. It shines light on some of the darkest times in history and the bad decisions that caused them. It also explains why current events are still being shaped by this persistent poison. This is the true story of lead arsenate pesticide and the coverup that continues to this day.

Lead arsenate is a compound that was used world-wide in enormous amounts as a pesticide on many food, cotton, rice and tobacco crops, as well as lawns and golf courses in the first half of the 1900's. Orchards that grew apples and oranges were some of the biggest users of this poison, which was mixed with water and sprayed directly onto the fruit trees several times a year. After aerial crop dusting airplanes were developed, around 1925, they were also used to spread the poisonous pesticide onto crops in both powder and liquid form.

Lead arsenate is a compound that contains both lead and arsenic. Both of these are toxic, and since they are elements they will never decompose into something that is nontoxic. **In the United States alone, about one billion pounds of lead arsenate was produced between 1925 and 1965.** The graph below shows that production increased steadily until around 1940, and was widely used until the 1960's when it was replaced with DDT and other pesticides. This trend was similar in many other countries.

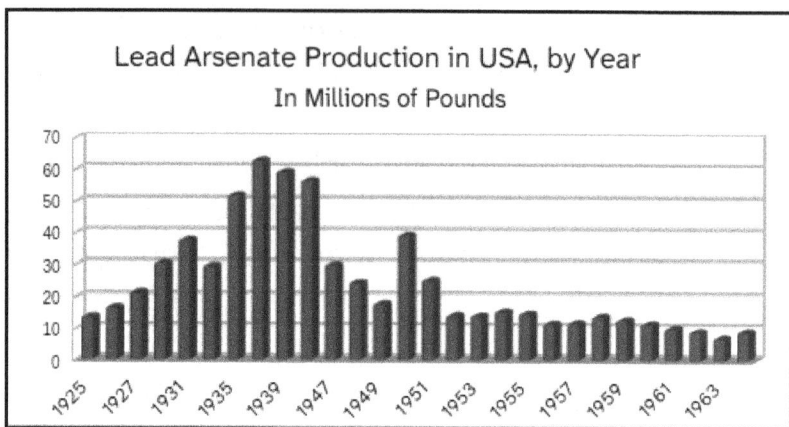

Figure 1 Lead Arsenate Production in USA.

The United States Department of Agriculture (USDA) produced yearly records of the amounts of various pesticides produced and used in the United States. These records only started in 1925, and were not readily available for the World War II period. For example, the figure 2 shows a typical record, table 817 from 1941, which was buried all the way back on page 701 of the USDA statistics book.[2] I looked at dozens of these records. Many other countries used similar amounts of lead arsenate, but since I only speak English I decided to concentrate my data collection from the United States records.

TABLE 817.—*Insecticides and fungicides: Production, sales, imports for consumption, and domestic exports, United States, 1934-39*

Principal items	1934	1935	1936	1937	1938	1939 [1]
Arsenic, white:	*Pounds*	*Pounds*	*Pounds*	*Pounds*	*Pounds*	*Pounds*
Production [3]	26,192,000	28,474,000	30,758,000	33,628,000	33,370,000	44,682,000
Sales: [3]						
Refined	13,186,000	11,370,000	13,652,000	13,466,000	7,464,000	10,738,000
Crude	18,060,000	13,970,000	17,510,000	21,805,000	18,856,000	34,140,000
Imports for consumption	28,219,797	30,149,938	35,171,964	38,511,324	28,475,783	29,347,810
Calcium arsenate:						
Production [4]		43,293,354		37,001,959		39,281,788
Imports for consumption	24,000	182,900	817,200	795,243	400,000	1,627,193
Exports	3,356,842	4,104,810	6,294,663	5,383,365	5,242,882	6,731,103
Lead arsenate:						
Production [4]		52,145,851		63,291,440		59,568,596
Imports for consumption				551		11,557
Exports	650,256	1,156,922	827,560	1,042,880	1,021,345	1,712,583
Paris green:						
Production [4]		2,638,210		1,834,340		2,040,307
Imports for consumption [3]	8,899	38,085	33,207	108,825	103,556	45,823

Figure 2: Table from USDA Agriculture Statistics 1941.

These records also document similar amounts of two other arsenic compounds, calcium arsenate, and copper arsenate which is also known as "Paris Green." A lethal dose of arsenic in humans is about one gram for an adult. **The amount of arsenic used on crops in the United States alone between 1925 and 1965, (about one billion pounds or 400 million kilograms) is enough doses to kill roughly 400 billion people, more than 1,000 times the current US population.**

The toxicity of lead has also been extensively studied due to the use of leaded gasoline that was sold in the past, but is now mostly banned around the world. These studies have shown that lead has been associated with lower intelligence, mental retardation, and an increased propensity to commit violent criminal acts. There is no arsenic in leaded gasoline, but the arsenic in lead arsenate also contributes to the toxic and behavioral effects of this pesticide.

This book presents the new hypothesis that the world-wide use of lead arsenate pesticide was the major factor in causing World War I, World War II, and the holocaust. I will also show how lead exposure from other sources could have been a major cause of wars, and other violent acts, earlier in history such as the during the Roman Empire and Viking raids.

This book will explain the unique chemical properties of lead that make it such a persistent poison, and describe how lead manages to stay in the environment for extremely long periods of time, contaminating our food supplies and well water. I will explore the great divide between people who drink well water contaminated with lead and those who get their water piped in from surface water supplies such as rivers and lakes that have much less lead, and are regularly tested to monitor their lead levels.

Later in this book I will explore the current effects of lead exposure and its relationship to violent crimes and political support for violent dictators and ethnic cleansing. I will propose the use of a new medical term, "Deranged Nazi Syndrome" (DNS), to describe a person who has been exposed to large amounts of lead and exhibits severe antisocial behavior especially towards people in other racial, ethnic, or social groups.

I will also present the hypothesis that the great political divide in the United States between the so-called red states and blue states is largely due to the amount of lead in their drinking water due to the source of their water (private wells vs. public water systems).

Figure 3: Mixing lead arsenate - 1945

Figure 4: Spraying Lead Arsenate 1924

STANDARD LEAD ARSENATE

AN AGRICULTURAL INSECTICIDE

ACTIVE INGREDIENT:
Standard Leaded Arsenate($PbHAsO_4$) 95%
INERT INGREDIENTS 5%
                                          100%

Lead expressed as Metallic 56.7%
Arsenic expressed as Metallic 20.5%
Arsenic in water soluble form expressed as
    Metallic, not more than 0.5%

☠ POISON ☠

ANTIDOTE: Give a tablespoonful of salt in a glass of warm water and repeat until vomit fluid is clear. Then give two tablespoonfuls of Epsom Salts or Milk of Magnesia in water, and plenty of milk and water. Have victim lie down and keep quiet. CALL A PHYSICIAN IMMEDIATELY!

DANGER!

Poisonous if swallowed. Avoid contact with skin, eyes, or clothing. Wash thoroughly after using. Avoid breathing dust or spray mist. Do not contaminate feed and food.

Keep Out of Reach of Children and Domestic Animals.

OBSERVE THE FOLLOWING USE PRECAUTIONS

In order that pesticidal residues on food and forage crops will not exceed tolerances established by the Federal Food and Drug Administration, use only at recommended rates and intervals, and no not apply closer to harvest than specified. Do not apply or allow to drift to areas occupied by unprotected humans or beneficial animals or onto adjoining food, fiber or pasture crops. The grower is responsible for residues on his crops as well as for damages caused by drift from his property to that of others.

Toxic to fish and wildlife. Do not contaminate any body of water or apply to any area not specified on the label.

NOTICE: Stauffer Chemical Company makes no warranty of merchantability or any other express or implied warranty concerning this material. It shall not be held responsible for personal injury, property damage, or other loss resulting from the handling, storage or use of this material. The buyer assumes all risk and liability resulting from such handling, storage or use.

To protect bees, do not apply when crops are in bloom. Remove or cover hives if application is necessary during bloom.

Do not use this product on feed or forage to be fed dairy animals or to livestock being finished for slaughter.

Consult State Agricultural Experiment Stations or the State Agricultural Extension Service for additional information, as the timing, number and rate of applications needed will vary with local conditions.

DIRECTIONS FOR USE

Measure the specified amount or this product then wash through filler screen in nearly filled spray tank or pre-mix in a bucket before pouring into spray tank. Add balance of water to fill tank. Keep agitator running during filling and spraying operations. Do not allow mixture to stand. Do not use in low volume gear pump spray equipment. Do not combine with emulsifiable liquids unless previous use of the mixture has proven physically compatible and safe to plants. Emulsifiable concentrates should always be thoroughly emulsified with at least half of total water before adding this product.

RECOMMENDATIONS

Rates are given in pounds of Standard Lead Arsenate per 100 gallons of water in full coverage sprays.
CHERRIES: Cherry Fruit Fly, Black Cherry Fly — Use 2½ to 3 lbs. plus 2½ to 3 lbs. hydrated lime. Begin all treatments when first adults are trapped. Make 2 or 3 applications at 10-14 day intervals.
Pear Slug — Use 2 lbs. plus 2 lbs. hydrated lime making 1 or 2 applications; the first about 20 days after bloom. Do not apply on cherries within 30 days of harvest. (May be applied up to 14 days of harvest on processing cherries if dosage rate limited to 24 lbs. per acre.) Dosage rate for English Morello variety should be reduced to 1 lbs. per 100 gal. water.
PEARS: Codling Moth — Use 3 lbs. Make 2 to 4 cover sprays beginning with calyx.
Pear Leaf Worm — Use 2 lbs. at petal fall
Pear Slug — Use 2 lbs. about two weeks after bloom.
Pear Curculio — Use 3 lbs. at petal fall and 10 days later. Do not apply within 30 days of harvest. Remove excess residue at time of harvest. DO NOT GRAZE LIVESTOCK IN TREATED GROVES.
USDA Reg. No. 476-374   S-SF-660511-WC

BURN BAG WHEN EMPTY — BURY ASHES

4 LBS. NET

STAUFFER CHEMICAL CO.
NEW YORK

Figure 5 Label from a bag of lead arsenate.

Chapter 2 – History of Lead Pesticide Use

Lead arsenate was a pesticide used world-wide in massive amounts for many decades in the late 1800's to the mid 1900's. It was used world-wide on many different types of crops, especially apple orchards, orange groves, fruit trees, and cotton fields. Lead arsenate ($PbHAsO_4$) was first used in 1892 as an insecticidal spray against the gypsy moth.

Apple production in the late 19th and early 20th centuries was very widespread. High production reflected localized agricultural markets and low fruit yield per acre. This contrasts sharply with today's world market and higher yields per acre. In 1925, for example, apples were grown on over 300,000 acres in Virginia (Taylor, 1926). Today Virginia farmers grow less than 12,000 acres of apples (Virginia Agricultural Statistics Service, 2006)[3].

Lead arsenate is a compound containing two heavy metals, lead and arsenic. Arsenate is the word used by chemists to denote a salt form of arsenic. Both lead and arsenic are known today to have very harmful and toxic properties, but in the past the long term effects of lead at lower doses was unknown.

Heavy metals like lead and arsenic are elements that are never destroyed. They do not break down over time like organic compounds, and they never evaporate. They also have low solubility in water, so they take an extremely long time to get washed away from rain water.

The lethal dose of arsenic in humans is 2-20 mg/kg, or 140 to 1400 mg (mg = milligrams) for an average-sized adult. A 140-mg potentially lethal dose is the same as 0.145 grams. Less than 1/8 teaspoon can be fatal to a healthy adult, while even less could kill a child, an adult with impaired health, or an elderly person.[4]

Figure 5 shows a label from a four pound bag of 95% pure lead arsenate. Note that lead makes up 57% by weight and arsenic makes up 20% . **Although arsenic was a well known poison, this four pound bag of pesticide contained enough lethal doses of arsenic to kill well over one thousand people.**

This four pound bag of pesticide cost about $1.40 in 1950 and one bag would treat around 12 apple trees over a year, or one fifth of an acre. This means that about 20 pounds of lead arsenate were used per acre every year.

If these orchards were sprayed the same way for 50 years, this means that about 1,000 pounds of lead arsenate were applied to every acre.

In 1930 the USDA recommended using 100 pounds of lead arsenate in a typical 3,000 square foot suburban lawn. **That is enough arsenic to kill over 20,000 people!**

Lead-arsenate treatment of lawns in suburban districts is a *practical method for controlling Japanese and Asiatic beetle grubs. Since the area of the average suburban lawn is less than 3,000 square feet, it would require less than 100 pounds of lead arsenate to poison the soil of a new lawn to a depth of 3 inches, and the cost should not exceed $15. This would protect the grass roots from injury by grubs for a period of four to five years. To top-dress an established lawn with lead arsenate and soil would cost proportionately less*

Walter E. Fleming,
Entomologist, Bureau of Entomology.[5]

Lead arsenate was also used on suburban lawns and golf courses at a recommended initial rate of 1500 pounds per acre to prevent bald spots from forming due to Japanese and Asiatic beetle grubs. This was effective for 4 or 5 years, but then additional applications were recommended for best results. Since a typical 18 hole golf course is about 150 acres in size, this recommendation means initially applying 225,000 pounds of lead arsenate to a 18 hole golf course, then reapplying nearly the same amount every 4 or 5 years. **That is enough arsenic to kill over 45,000,000 people!   That's 45 million lethal doses of arsenic applied to every 18 hole golf course every 4 or 5 years!**

The arsenic from just three golf courses treated this way in 1945 would have contained enough poison to kill everyone living in America at that time (about 140 million people).

Chapter 3 – Lead is Forever

Lead arsenate is a chemical compound that contains two heavy metals, lead and arsenic. Both of these are basic chemical elements that are never destroyed. Even though the use of lead arsenate was greatly reduced or even eliminated in most countries more than 50 years ago, both of these heavy metal elements are still around. They are simply never broken down into something other than lead and arsenic.

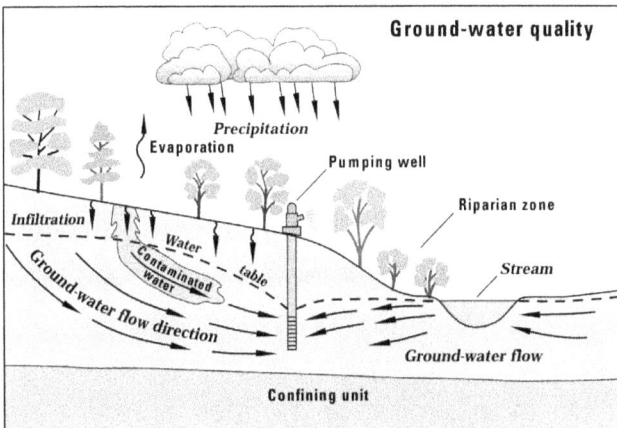

Figure 6: Ground water contamination from farms.[6]

The low water solubility of lead, and to a lesser extent arsenic, also causes these heavy metals to continue to persist in the areas that they were applied. They can't simply wash away or decompose like most modern pesticides. The low water solubility of lead means that the lead will slowly leach into our underground aquifers and well-water supplies, contaminating them for centuries.

The above figure shows a simple graphic of how lead arsenate leaches from the top soil of farms into the water table and then by ground water flow into well water. The contaminated water not only contaminates the drinking water of households that are supplied by wells, but it also contaminates the water used on farms for livestock and irrigation of crops. Contaminated livestock and crops are another pathway leading to lead arsenate ingestion by humans.

On farms, most of the lead arsenate that doesn't enter the food supply is re-released into the top soil of farms by animal urination and crop wastes being tilled into the top soil. A small portion of the lead arsenate from one farm may also enter local streams which might be used on other farms or supply drinking water for municipal water systems. Any lead arsenate that enters a stream and makes it all the way to the ocean is effectively removed from any substantial human or animal consumption.

Lead compounds, such as lead arsenate, generally have two chemical properties that greatly affect how long they persist in contaminated areas. These properties are the low water solubility and the heavy weight of most lead compounds.

Direct measurements of the amount of lead and arsenic in soil can be made but these tests are expensive and multiple tests are needed in each test area as the distribution of lead and arsenic in soil is not very uniform. In today's contaminated soils these types of tests generally give results in the the hundreds of mg/kg for lead, and 50 to 200 mg/kg for arsenic.

To put this in more understandable terms, every shovel full of contaminated soil contains enough arsenic to kill a

person, as well as enough lead to kill a person.

This lead and arsenic will slowly leach out of the soil with every rainfall and much of it will end up in the underground water aquifers that are the source of well-water.

The low solubility of lead also contributes to errors and inaccuracies in scientific tests that are used to determine the amount of lead in drinking water supplies. To illustrate this problem try to imagine that you have a large barrel in which you placed 50 pounds of rusty iron nails and 50 pounds of sand. The iron and the sand represent the soil of our farmland, and the water represents underground water aquifers. The best way is to determine how much iron is in the sand is to find out how much iron was originally put in the barrel. This is analogous to how I determined the amount of lead arsenate that was placed on American farmland, lawns, and golf courses.

A vastly inferior way to determine the amount of iron in the barrel is to poke holes in the bottom and run water into the top and measure how much iron, as rust, is in the water that seeps out the bottom. This is analogous to the methods currently used to measure lead in potable water supplies such as wells and ground aquifers. The problem with this type of measurement is that iron nails are not soluble and they need to rust before any iron rust comes out the bottom. The process of rusting does not occur at a constant rate and is highly dependent on other factors such as the pH of the water and the presence of salts in the water, as well as how long the iron nails were wet or dry.

Another problem with this method is that rust is not very soluble in water and more of it can be suspended in water by stirring or agitation. If it is left to rest most of the iron rust will precipitate to the bottom, and if a sample is taken the amount or iron will vary depending on where the sample was taken. The amount of iron rust will also vary greatly with the rate that water is added to, and extracted from, the barrel.

This second method is analogous to most of the types of lead assays that are performed on potable water supplies. This is partly because it is usually mandated by law or other regulations, and that the first method is generally impractical on a large scale. However, the second method can give wildly different results and can be manipulated to give smaller amounts.

For example, many testing procedures call for running tap water for 10 minutes before taking a water sample. This leads to smaller amounts of lead being detected. In the rusty nails in a barrel analogy it would be like turning the hose on high and discarding the first 10 minutes worth of water eluting out of the bottom of the barrel, which would be very rusty. After 10 minutes of rinsing the rust out with water most of the readily soluble or suspended iron rust will have been washed away which gives us a sample that is much lower in iron and doesn't correlate well with the amount of iron contained in the barrel.

If we apply this analogy to the amount of lead in underground water supplies we can see that measuring lead in tap water using current procedures is probably going to vastly underestimate the amount of lead in groundwater aquifers, along with the amount of lead consumed by humans and livestock. But at least it gives us some data which is useful for comparing the relative amounts from different areas and water sources.

Lead Use and Ground Water Aquifers

During the early 1900's so much lead arsenate was used worldwide on crops and lawns that it had an immediate violent effect on the people who ate the crops, such as the Nazi's. This will be explored in much more detail later in this book.

However, over the past century there has been so much lead and arsenic slowly seeping into underground water aquifers that these are dangerously polluted with lead and arsenic. These ground water aquifers are often the sole source of water in many rural areas. Some of these aquifers span hundreds of miles transporting lead and arsenic from lawns and croplands into other drier areas that are not suitable for lawns or orchard crops.

The water from these lead and arsenic polluted aquifers are often pumped from private and municipal wells for both human consumption and to water crops and livestock. Because of this mechanism, humans, crops, and livestock in many areas are currently being exposed to dangerous levels of lead and arsenic. However, this type of lead exposure depends on well-water use and varies greatly by location (distance from the pesticide application).

In most parts of the world, municipal water systems that draw their water from surface water supplies, such as rivers, will have much lower lead and arsenic than water that originates from wells. This is especially true for private home well-water systems that are often unregulated and untested for lead and other pollutants.

In the next several chapters of this book I will explore the connection between lead exposure and propensity for violence as well as the decreases in intelligence that are associated with lead exposure. After that I will explore the geographical relationships between well-water use and lead exposure with the intellectual and behavioral changes caused by lead exposure.

Chapter 4 – Lead and violence in Ancient Times

The senseless violence of the Roman Empire is very well known today even though it occurred around 2000 years ago. Most Roman cities built expensive coliseums mainly for gladiator fights. These bloody fights to the death were well attended by both the citizens and rulers of the area who seemed to enjoy watching people and animals being bloodily beaten to death. Roman emperors and military officers routinely ordered mass executions of prisoners and even performed crucifixions along roadways so that people would see the slow painful death that awaited anyone who dared fight against the Roman Empire.

The ancient Romans also used a lot a lead. It was widely used in plumbing pipes in historic Rome, and was added to wine to preserve it and give it a better taste. In hindsight it seems reasonable to associate Roman lead exposure to the violence and xenophobia that it probably caused.

Figure 7: Pollice Verso ("With a Turned Thumb"), an 1872 painting by Jean-Léon Gérôme

It is estimated that the number of workers who were occupationally exposed to lead during the period of the Roman Empire was over 140,000 per year. Considerably higher fractions of lead-using populations were exposed to lead contamination in their food and drink. The Romans, for example, preserved their fruits and vegetables with lead salts, cooked their foods in leaden pots, and commonly assuaged their "sweet teeth" with the sugar of lead (saccharum saturni, or lead acetate). They added lead to their wines to stop further fermentation, to impart color or bouquet, or to blunt the acidity of an erratic brew. Their water was delivered in lead pipes, while saturnine [lead based] cosmetics and medicaments were common and quite popular. With such overexposure to lead, we find frequent literary references to epidemics of plumbism and saturnine gout among the members of the Roman aristocracy. Other historiographic evidence indicating that lead poisoning caused the reproductive failure of the ruling oligarchy includes the high incidence of sterility, the alarming rates of stillbirths, and the well-known mental incompetence of the progeny of the aristocrats. Indeed, the psychological profiles of the emperors and usurpers who reigned between 50 B.C. and 250 A.D. suggest that the majority of *them probably suffered from lead poisoning.*

Lead and Lead Poisoning in Antiquity
Jerome O. Nriagu 1983 [7]

The Vikings were another ancient population that also seems to have a large number of violent episodes against innocent people, such as the Viking raids that they unleashed against English and European coastal towns. There is some evidence to suggest that many Vikings were exposed to high levels of lead due to the mining and blacksmithing techniques that they used to make tools and weapons.

Another irrational and violent episode from the past, the Salem witch trials, which occurred around 1692 may have been caused by lead in their environment. Lead was mined in areas near Salem Massachusetts (USA) in more recent times, suggesting that the land was rich in lead.

Figure 8 Viking swords.

Chapter 5 – Symptoms of Lead Poisoning

Lead poisoning can have many different symptoms in humans. Some possible symptoms of long-term lead exposure are:

1) Pale skin
2) Stunted bone growth
3) Hunchback (Kyphosis) – due to stunted bone growth
4) Eyes that drift up (oculogyric crisis)
5) Difficulty controlling handwriting size (such as micrographia or macrographia)
6) Mental retardation
7) Increased Propensity for violence
8) Xenophobia (fear or hatred of people from other races or cultures)
9) Involuntary repetition of syllables, words, or phrases (palilalia)
10) Encephalitis lethargica symptoms (a devastating epidemic around 1915 to 1926)
11) Parkinsonism or Post-encephalitic Parkinsonism
12) Increased blood pressure

These symptoms will be discussed in much more detail later in this book, but for now I would like to concentrate on just two of these symptoms, mental retardation and increased propensity for violence. I have already talked about the violence in the Roman Empire which was probably caused by lead poisoning. In the next chapter I will discuss modern age mental retardation and violence in polluted urban areas caused by the use of leaded gasoline.

In 2022 The World Health Organization (WHO) published a report claiming that:

Exposure to lead is associated with a wide range of effects, including various neurodevelopmental effects, mortality (mainly due to cardiovascular diseases), impaired renal function, hypertension, impaired fertility and adverse pregnancy outcomes. Impaired neurodevelopment in children is generally associated with lower blood lead concentrations than the other effects, the weight of evidence is greater for neurodevelopmental effects than for other health effects and the results across studies are more consistent than those for other effects. For adults, the adverse effect associated with lowest blood lead concentrations for which the weight of evidence is greatest and most consistent is a lead-associated increase in systolic blood pressure.

The WHO also said that lead levels in blood at concentrations of *25 µg/kg body weight* [are] *associated with a decrease of at least 3 intelligence quotient (IQ) points in children and an increase in systolic blood pressure of approximately 3 mmHg (0.4 kPa) in adults. These changes are important when viewed as a shift in the distribution of IQ or blood pressure within a population.* [8]

There are many other scientific papers exploring the link between decreased IQ and increased violence in lead-polluted urban areas all around the world. The really striking evidence is based on the fact that the ban on the sale of leaded gasoline occurred at different times in different countries. Around the world, nearly all the various lead-polluted urban areas have seen a increase in IQ and a dramatic decrease in violence within a few years after the banning the sale of leaded gasoline, even though they occurred in different decades.

Chapter 6 – Leaded Gasoline's Effect on IQ and Violence in Urban Areas

Much of the data linking lead to violent behavior comes from studies that examine the links between the use of large amounts of leaded gasoline and the crime and violence that increased because of it. The amount of lead added to gasoline was enormous. In the USA alone, over 5 million pounds a week were added to gasoline. Most of this lead was exhausted out of tailpipes as extremely fine metallic particulates that can be absorbed by breathing. [9]

There are many scientific papers written about the connection between leaded gasoline, IQ, and propensity to violence. These can be easily found online by searching PubMed at https://pubmed.ncbi.nlm.nih.gov/. An excellent example of this type of research is a paper titled *Environmental causes of violence* by D.O. Carpenter and R. Nevin in

Physiology & Behavior[10] The next quote and two figures are from this paper.

> *Violent and anti-social behavior is usually attributed to social factors, including poverty, poor education, and family instability. There is also evidence that many forms of violent behavior are more frequent in individuals of lower IQ. The role of exposure to environmental contaminants has received little attention as a factor predisposing to violent behavior. However a number of environmental exposures are documented to result in a common pattern of neurobehavioral effects, including lowered IQ, shortened attention span, and increased frequency of antisocial behavior. This pattern is best described for children exposed to lead early in life, but a similar pattern is seen upon exposure to polychlorinated biphenyls and methyl mercury. Although not as extensively studied, similar decrements in IQ*

are seen upon exposure to arsenic and secondhand smoke (SHS) exposure. Prenatal and postnatal SHS exposure is also associated with increased rates of conduct disorder and attention deficit hyperactivity. Recent evidence suggests that temporal trends in rates of violent crime in many nations are consistent with earlier preschool blood lead trends, with a lag of about 20 years. These ecologic correlations are consistent with many controlled studies suggesting that lead-exposed children suffer irreversible brain alterations that make them more likely to commit violent crimes as young adults. If this pattern is true for lead and other contaminants, the most effective way to fight crime may be to prevent exposure to these contaminants.

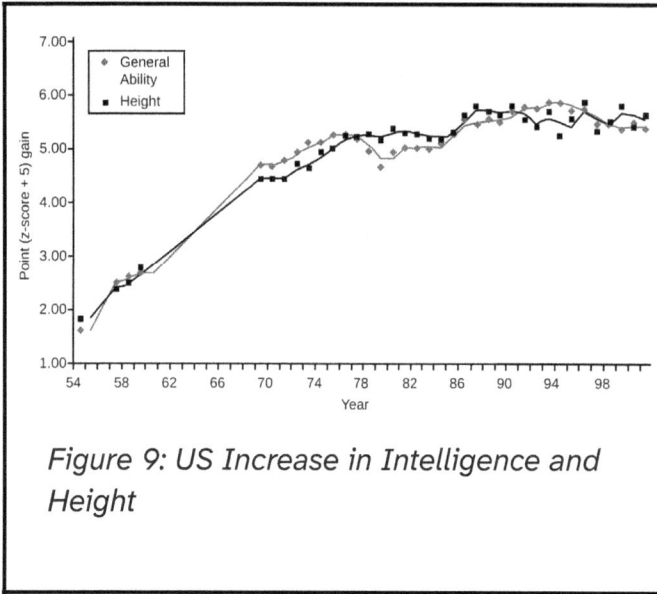

Figure 9: US Increase in Intelligence and Height

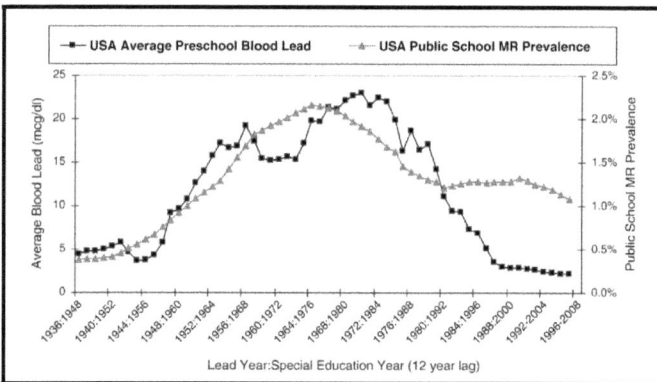

Figure 10: USA public school mental retardation (MR) prevalence and preschool blood lead trends. The rise and fall of preschool blood lead from 1936 to 1990 explains 65% of the substantial 1948-2001

variation in the percent of public school students in special education for mental retardation (MR). The 12-year time lag for students ages 6–18 is consistent with lead-induced cognitive damage in the first year of life.

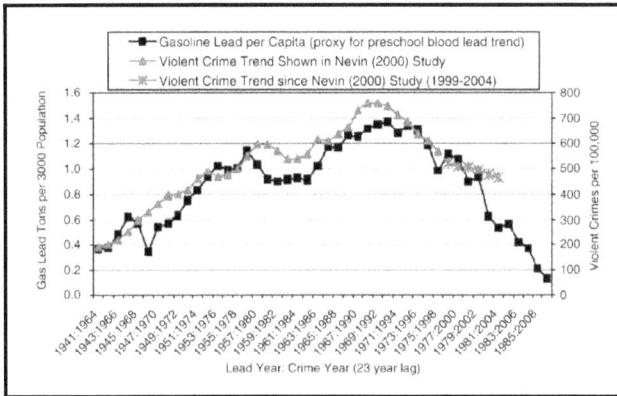

Figure 11: USA violent crime and lead exposure trends and age-specific violent crime arrest rate shifts. In the upper plot gasoline lead per capita (left axis) and violent crime rate per 100,000 (right axis) are plotted against time, with the crime rate shifted for a 23 year lag (data from 55). The lower plot shows age-specific violent crime arrest rates (data from the Office of Juvenile Justice and Delinquency Prevention, 2004) and shows that the peak offending age shifted to older ages by 2001, as the 1990s violent crime decline

*was associated with an especially sharp
decline among juveniles born after the
early 1980s decline in gas lead levels.*

Since cars are more dense in urban areas,
they emitted much more lead in urban areas.
And since lead can cause decreases in
intelligence, and increase the propensity for
violence, these effects were greater in urban
areas. In the USA many large cities have
large populations of Black people and they
were disproportionally affected by the lead
from car exhaust. (It may also be true that
Black people were more affected by lead
arsenate pesticide spraying, but solid
evidence for this is sparse.)

Since Black people were more affected by
lead exhaust fumes, this gave rise to false
racist stereotypes about Black people being
genetically inferior to other races. For
example, the book *The Bell Curve:
Intelligence and Class Structure in American
Life*[11] (Free Press, 1994) uses a lot of
statistical correlations to falsely suggest that
Black people were genetically inferior in
intelligence and propensity to violence.
(Donald Trump seems to believe this as he

regularly makes references to Black people he dislikes as "low IQ individuals".)

Every competent scientist knows that **correlation is not causation**. The high correlation between the number of people using umbrellas and rainfall amount should not lead to the conclusion that umbrellas cause rainfall.

The high correlation between pain reliever use and hospitalization does not mean that the pain reliever caused the hospitalization. It is more likely that an illness caused pain which made the person take a pain reliever. When this didn't relieve the pain the person went to the hospital.

The effects of lead exposure in urban centers from leaded gasoline occurred world-wide. Many other cities around the world had similar effects from leaded gasoline, but their urban centers were not predominately Black, for example Mexico City. It was the lead from leaded gasoline that was causing these effects, not racial genetics.

Although leaded gasoline was largely introduced into the world at the same time, with the advent of the automobile, the decline in leaded gasoline was generally due to laws prohibiting it. Some countries banned leaded gasoline in the 1970's and 1980's while others waited until decades later. Many of these studies show very close correlation between the banning of leaded gasoline and a decrease in crime and violence, especially in urban areas where lead exposure is the greatest.

A review paper had this to say about lead and violence:

> *Prenatal and early childhood lead exposure was reported in correlation with violent crimes in adulthood (Park et al., 2008). The highest lead levels in the air were also shown to deviate normal behaviour and turn to become aggressive and violent, thus for example the highest murder rates were found in countries with high levels of lead in the air (Neelemann et al., 2004).* **One**

study theorizes that lead exposure explains 65% to 90% of the variation in violent crime rates in the US *(Shih et al., 2007; Kosnett et al., 2007). Another study showed a strong association between preschool blood lead levels and subsequent crime rate trends over several decades across nine countries (Nevin, 2007).*[12]

Lead exposure from leaded gasoline would tend to be from inhaling very small particles of lead generated by engine exhaust. The amount of lead absorbed in this manner can be expected to substantially decrease within a decade after the banning of leaded gasoline in that area.

Lead exposure from lead arsenate pesticides acts in a different manner. Much of the initial exposure to the general public probably occurred orally through the ingestion of food that was sprayed with lead arsenate. Then lead arsenate contaminated the soils and waters under the farms where it was used.

The low water solubility of lead, and to a lesser extent arsenic, also causes these heavy metals to continue to persist in the areas where they were applied. They can't simply wash away or decompose like most modern pesticides. The low water solubility of lead means that the lead will slowly leach into our underground aquifers and well-water supplies, contaminating them for centuries.

Chapter 7 – Lead and The Flynn Effect

"The Flynn effect" is a term used to describe the large increase in intelligence that were measured in many parts of the world over the 20th century, named after researcher James Flynn (1934-2020). Most of these intelligence gains occurred after the 1950's when the use of lead arsenate declined due to better pesticides becoming available.

For example, Dutch military conscripts gained 21 IQ points during only 30 years, or 7 points per decade, between 1952 and 1982.[13]

The following figure shows the Flynn effect increase in intelligence in the United States and the increase in height. Since lead is widely believed to stunt bone growth, the extremely close correlation between height and intelligence point to the decrease in environmental lead as the cause of this increase in intelligence. However, the role of better nutrition may also play a substantial role and can not be ruled out.

The next figure shows the Flynn effect increases in IQ for several different parts of the world.

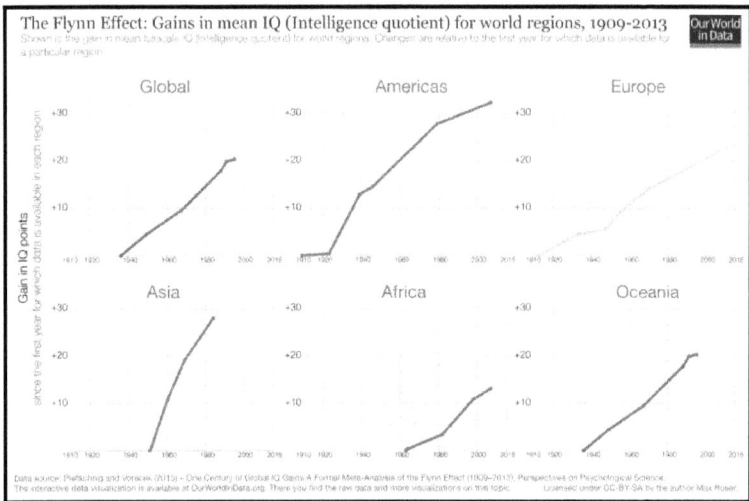

The Flynn Effect: Gains in mean IQ (Intelligence quotient) for world regions, 1909-2013

However, these gains in intelligence were preceded by by a general loss of intelligence that occurred in the early 1900's that is not show in the above graphs.

IQ (intelligence quotient) testing was taken up enthusiastically in the US. Early versions had already produced some interesting results – in 1915 the mayor of Chicago tested as a "moron" (the newly invented term for a "feeble minded" person) on one version of the Binet [IQ] scales.

In 1917, Harvard psychologist Robert M Yerkes decided that the world war, which involved the mass mobilisation of soldiers into the US army, presented an unmissable opportunity to demonstrate the IQ test's scientific validity.

Drafted into the army as a colonel, he presided over the administration of mental tests to 1.75 million men, something he called a feat of "human engineering". Tests were administered in written form for the literate and in pictorial form for those who could not read. They were timed and often taken simultaneously by large numbers of men under the supervision of an examiner in specially commissioned buildings.

The results came as something of a shock. According to the data, the average white American had a mental age of 13. This stood just marginally above the designated mental age for a moron, which was between eight and 12. Furthermore, the results suggested that 37% of white Americans and 89% of "negroes" fell within this category.

This all took some explaining. It confirmed many of the racial and class prejudices of Yerkes and his fellow psychologists, who were all hereditarians, believing that intelligence or a lack of it was inherited and lowest among non-whites and the poor of all races.

However, the sheer scale of it seemed to suggest the whole country was rapidly degenerating, with even the white race – God forbid – heading for mental oblivion.

The great American IQ panic of 1917[14]
Community Living Magazine 21 April 2018 — Spring 2018

This magazine article continues with various explanations of the flaws of the IQ tests of that era. And while there were many valid sources of errors, it seems impossible to me that the massive changes in measured IQ were solely due to problems with the test.

I believe that the widespread decreases in intelligence during the first half of the 1900's was due to the widespread use of lead arsenate pesticides during that time. I also believe that the increase in intelligence in the later half of the 1900's (the Flynn effect) was largely due to the decreasing use of lead arsenate pesticide and leaded gasoline that started during that period of time.

Chapter 8 – Hitler's Symptoms

In the early twentieth century, before World War II, much of Europe and the developed world was devastated by lead arsenate and violence. Spain suffered through a civil war which was noted for its extreme violence. Armenians were slaughtered in a genocide during World War I. People were routinely going to political rallies which often turned into deadly street riots. A massive increase in violent serial killers and mass murderers occurred.

America and England were not spared from this violence and xenophobia. The Ku Klux Klan (KKK), an extremely violent and racist terrorist group, had a vast increase in membership and even held mass marches in Washington DC. Several groups of Nazi supporters were growing in size in both the US and England right up to the start of World War II, and then were largely eradicated.

Then World War II and the Holocaust occurred with so much violence and evil behavior that it still shocks the world, and leaves rational people wondering how this could have ever happened in such an educated society.

Similar irrational and extreme violence was inflicted by Japanese soldiers on civilians as well as soldiers. Hundreds of thousands of civilians and soldiers were murdered in the Rape of Nanking (also known as Nanjing Massacre). There are many accounts of Chinese women and young girls being gang-raped by Japanese soldiers then murdered when they were done with them.

The Japanese sometimes talk of "The Winds of War" which overtook their country. Many people in Japan and the rest of the world struggle in understand how "The Winds of War" caused the Japanese people to engage in such violent and immoral behavior.

Today many people, including myself, struggle to make sense of the causes of all this extreme violence and xenophobia. (Xenophobia means the fear and hatred of strangers or foreigners.) An extremely large number of books and articles have been written by historians and psychologists seeking to explain how these generally well-educated and well-developed societies evolved into a grotesque murderous orgy of violence and xenophobia. Many of these authors have suggested several different factors to explain why much of the world exploded in violence in the 20th century, but to the best of my knowledge none of them have explored the connection between the use of lead arsenate pesticide and the massive increase in violence and xenophobia.

One of the main themes of this book is to show a causal link between the use of lead arsenate pesticide in food stocks and this historical violence. Another similar theme is to show the current connection between drinking well water contaminated by past lead arsenate pesticide use and an increase in xenophobia and senseless xenophobic violence in the 21st century.

America has seen a massive increase in random xenophobic mass shootings in the

21st century, as well as vehicles being used in mass killings at large gatherings such as parades and street festivals. Europe also has similar troubles.

In order to understand this modern day second wave of lead-induced "Deranged Nazi Syndrome" due to lead arsenate in drinking water, it is important to understand the first wave of "Deranged Nazi Syndrome" that was probably due to lead arsenate exposure from food in the first half of the 20th century.

Adolph Hitler was known to be a vegetarian, and since he didn't eat meat it seems reasonable to suggest that he probably ate more fruits and vegetables to make up for the calories he wasn't getting from meat. The next figure is a photograph of Hitler at an outdoor banquet with some military officers sitting at the tables while lower ranked soldiers posed in the background. A very large serving plate loaded up with apples and other fruits was placed directly in front of him. These facts seem to suggest that Hitler probably ate a large quantity of apples and other fruits. Since apples and other fruit trees were sprayed with large quantities of lead arsenate, it is reasonable to suggest that Hitler was probably exposed to more lead arsenate than other non-vegetarian Germans.

It's not too important to know whether Hitler had more lead arsenate exposure than the rest of the German People. It is probably true that nearly all the people of Germany, Europe, North America, and the rest of the developed world were exposed to massive amounts of lead arsenate through their food supply. However, because of the vast amount of documentation about Hitler, he serves as a good subject to document the physical and mental effects of lead arsenate exposure.

Figure 12: Hitler eating fruit.

Many well educated people have written papers and books trying to explain why Hitler acted the way he did. Most of these fall well short of a definitive answer. Many of Adolph Hitler's medical conditions and symptoms are fairly well documented and point to lead arsenate exposure as the primary cause of both his physical and mental impairments. I will examine Hitler as the ultimate example of "Deranged Nazi Syndrome" and explain how lead arsenate exposure is linked to both his physical and mental states.

Some possible symptoms of long-term lead exposure are:

1) Pale skin
2) Stunted bone growth
3) Hunchback (Kyphosis) – due to stunted bone growth
4) Eyes that drift up (oculogyric crisis)
5) Difficulty controlling handwriting size (such as micrographia or macrographia)
6) Mental retardation
7) Increased Propensity for violence
8) Xenophobia
9) Involuntary repetition of syllables, words, or phrases (palilalia)
10) Encephalitis lethargica symptoms (a devastating epidemic around 1915 to 1926)
11) Parkinsonism or Post-encephalitic Parkinsonism
12) Increased blood pressure

It is important to note that all of these symptoms may not be present in individual cases of lead exposure, and these symptoms may have other causes. For example, pale skin may be a symptom of lead exposure, but it is often hard to tell if skin is pale due to differences in sun exposure and ethnic background.

A hunchback (kyphosis) may be due to lead exposure which possibly affects bone growth in the spinal cord. The next figure shows the strong difference in the angle of the neck and how it causes the head to tilt down. Hitler showed evidence of this towards the end of his life.

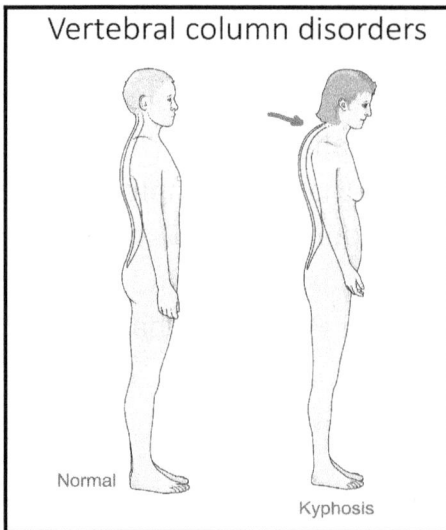

Figure 13: Kyphosis (hunchback).
Modified from a Wikipedia diagram.

Another possible symptom of lead exposure are eyes that drift up fairly often (an oculogyric crisis). This may be hard to notice in people with a hunchback because they tend to tilt their head down and when their eyes drift up they are looking straight ahead. There are some reports of Hitler having this symptom towards the end of his life.

Figure 14: Hitler's Handwriting Changes With Age (Micrographia).

Difficulty controlling handwriting size (such as micrographia or macrographia) is another possible symptom of lead exposure. Hitler had handwriting that became much smaller throughout his life (micrographia)[15], but others can exhibit handwriting that becomes much larger and composed of simple long strokes (macrographia). Both of these symptoms have the same root cause, a lack of fine motor skills.

Figure 15: The Very Large and Simple Strokes of Donald Trump's Signature Are Signs of Macrographia, and His Neck Shows Signs of Kyphosis (Hunchback).

Mental retardation is another symptom of severe lead exposure. Milder mental deficits have been associated with large urban areas that were exposed to car exhaust fumes when lead was used in gasoline. Different countries banned leaded gasoline at different times, and there is a close correlation between the time when leaded gasoline was banned in an area and the rebound in average intelligence in those areas.

The lead exposure from leaded gasoline is also associated with an increased propensity for violence. All around the world, violence increased in urban areas with the increased use of leaded gasoline, and there was a remarkable decrease in urban violence that occurred when leaded gasoline was banned in a particular area.

Xenophobia is currently not well documented as a symptom of lead exposure. But since there is firm evidence linking lead to mental retardation and propensity for violence, and these two symptoms are often associated with xenophobic behaviors, it is reasonable to suggest that lead exposure is linked to xenophobic behavior, especially violent xenophobic behavior such as the Holocaust and racially motivated mass shootings.

Xenophobia caused by lead exposure usually starts as a mild dislike of people from other races and cultures. If the lead exposure is high the symptoms may increase in intensity to hatred of people from other races and cultures as well as anyone who doesn't share your hatred. This can progress to taking physical actions against these people that can result in great harm such as randomly shooting them.

In severe cases of lead exposure that affect large populations, the large number of individuals affected can have a synergistic effect that increases the symptoms of the whole population. This can result in mob violence and supporting politicians who share their vile and disgusting xenophobic views.

When politicians and governments get involved in this lead-induced hatred the effects can spiral out of control. In Nazi Germany many people supported the deportation of Jews, and other undesirable people, out of Germany. Many of these Germans also supported placing them out of sight in dirty prison "work camps" where they wouldn't be seen again. Many Germans knew, or suspected, that these victims were never going to make it out of these concentration camps, but they didn't care. They felt that they were justified in defending their culture and their country from these "foreign invaders".

Today we are experiencing the second wave of Deranged Nazi Syndrome which was brought on by lead arsenate slowly seeping into our well water supplies, but these vile and disgusting xenophobic symptoms are the same.

Many of the same countries that were affected by the first wave of lead-induced violence are now putting up barriers against immigration by calling immigrants poor dirty criminals who are invading their country. Many of these countries are behaving just like Nazi Germany by deporting immigrants to foreign prisons where they won't be seen or heard from again, and there is little concern for their health or safety, or whether they live or die.

The time will come when today's xenophobic Nazi's will be executed for their crimes against humanity. These brain damaged lead-heads are probably too far gone for rehabilitation or redemption, and they will need to be executed to prevent any further harm to society. As the saying goes, The only good Nazi is a dead Nazi.

Figure 16: German Concentration Camp in Poland 1945

Figure 17: US Concentration Camp in El Salvador 2025.

Encephalitis lethargica was a devastating disease of previously unknown origin that spread around the world around 1915 to 1930. It is estimated that over a million people were affected and half a million died.

Encephalitis lethargica is characterized by high fever, sore throat, headache, lethargy, double vision, delayed physical and mental response, sleep inversion and catatonia. In severe cases, patients may enter a coma-like state (akinetic mutism). Patients may also experience abnormal eye movements ("oculogyric crises"), Parkinsonism, upper body weakness, muscular pains, tremors, neck rigidity, and behavioral changes including psychosis. Klazomania, a vocal tic involving compulsive screaming, is sometimes present.

Approximately a third of the affected children experienced change of behavior, with many of them becoming "delinquents". Boys between the ages of 5 and 18 years were the most affected. Symptoms include change of personality, restlessness, irregular sleeping habits, emotional instability manifesting as irritability, crying spells, and temper tantrums, including impulsivity, and unpredictability, what Economo described as "moral insanity". More extreme cases include aggression and "shameless sexual activity". Children under the age of 5 years suffered severe developmental delays. Delays were also present in children between 5 and 14 years of age, even though the claims are controversial.

It is estimated that 25-90% of adults also suffered from psychological problems, including hysteria and abnormal behavior and movement. A large minority of patients described having bradyphrenia.

--From Wikipedia article on *"Encephalitis lethargica"* Retrieved 7-14-2025.

Bradyphrenia is the slowness of thought common to many disorders of the brain. Disorders characterized by bradyphrenia include Parkinson's disease and forms of schizophrenia consequently causing a delayed response and fatigue. Patients with bradyphrenia may describe or may manifest slowed thought processes, evidenced by increased latency of response and also involve severe memory impairment and poor motor control. The word 'bradyphrenia' originates from the ancient Greek meaning 'slow mind.'

--From   Wikipedia   article   on *"Bradyphrenia"*   Retrieved   7-14-2025.

The plot of *Awakenings* (1990), an American movie inspired by real events, starring Robin Williams and Robert De Nero, revolved around a doctor who searched for a cure for encephalitis lethargica. This movie was nominated for three Academy   Awards,   and   has   some   good characterizations of patients who were severely affected by encephalitis lethargica.

Technical Notes:

Parkinson's _Syndrome_ or _Parkinsonism_ is a   medical   syndrome   characterized   by several specific clinical signs. Some people may   be   familiar   with   the   symptoms   of *Parkinson's Disease* due to the efforts of another   famous   actor,   Michael   J.   Fox. However,   the   term   *Parkinson's Disease* refers to a specific disease that may not be caused by lead exposure.

Parkinsonism or *Parkinson's Syndrome* refers only to a certain group of symptoms, some of which may also be present in patients with *Parkinson's Disease*. However, I am mostly concerned about another disease, *post-encephalitic Parkinsonism*, which generally occurs after a patient had *encephalitis lethargica*, which was described above.

I suspect that *post-encephalitic Parkinsonism* is due to the lingering effects of lead exposure which can cause *encephalitis lethargica* which can then lead to *post-encephalitic Parkinsonism*. Medical researchers may want to investigate this link and see if it has any relationship to *Parkinson's Disease*. However, since this book is intended for general audiences I will not be going into too much detail.

Hitler's Symptoms

Many of Adolph Hitler's medical symptoms suggest that he was suffering from lead poisoning. Note the similarities between the lead poisoning symptoms I listed above with some of the symptoms that Adolph Hitler was exhibiting.

Research works have suggested almost incontrovertibly, that Adolf Hitler suffered from Parkinsonism. However, the precise nature of his illness had always been controversial and post-encephalitic and idiopathic varieties were the ones which were most commonly thought as the possible etiology. He displayed features like oculogyric crisis, palilalia, and autonomic symptoms which strongly implicate post-encephalitic etiology in the genesis of his illness. Others on the contrary, observed premorbid personality traits like non-flinching mental rigidity, extreme inflexibility, and awesome pedantry; which are often observed in idiopathic Parkinson's disease. Moreover, nonmotor symptoms like disturbed sleep, proneness to temper tantrums, phases of depression, suspiciousness, and lack of trust on colleagues have also been described by various authors. Additionally, he was prescribed methamphetamine by his personal doctor and that might have led to the development of some of the later traits in his personality.[16]

Adolf Hitler and His Parkinsonism
Kalyan B. Bhattacharyya;
Ann Indian Acad Neurol
2015;18:387-390

The similarity of the symptoms quoted above for Hitler match up fairly nicely to the symptoms I previously listed for lead exposure.

Donald Trump's Symptoms

Donald Trump also appears to have many of the same symptoms related to lead poisoning. I believe that he has the first eight items on the list above. He has very pale skin underneath the facial bronzer that he uses. He also appears to have kyphosis (hunchback) and eyes that drift up (oculogyric crisis).

He clearly has macrographia (large handwriting) and suffers from mild mental retardation to some extent. He also has xenophobia and may have a propensity to support violent actions, although he seems to use other people to perform the actual violence, such as trying to kill his own Vice-President in order to overturn an election that they lost.

Chapter 9 – Lead is Linked to Parkinson's Disease by Well Water

There are many scientific studies that have seen an increased risk of getting Parkinson's Disease for people who live in houses that use well water compared to houses using public water supplies. The theory is that most private wells are not tested for ground water contamination, while public water supplies are regularly tested, and often use surface water supplies like rivers instead of wells.

A different study from researchers using a large database from the Mayo Clinic Rochester Epidemiology Project shows that living near golf courses could increase the risk of Parkinson's Disease.

Overall, the analysis revealed that people who lived within one mile of a golf course were 126% (or 2.26 times) more likely to receive a Parkinson's diagnosis than those whose homes were six or more miles away. Being farther from the fairway seemed to help; risk steadily tapered off beyond one mile, with the odds of PD diagnoses decreasing by 9% for each mile of distance from a golf course.

Distance is only part of the story. When researchers looked at households served by a public water system that contained at least one golf course, Parkinson's risk was 96% higher compared to households whose water systems did not have a golf course within their boundaries, and about 50% higher than people who use private wells. Additionally, when a golf course was in an area with groundwater vulnerable to contamination, the risk of Parkinson's was 82% higher than in less vulnerable areas with a golf course.

Taken together, the findings suggest that the pesticides and herbicides used to keep putting greens immaculate may be leaching into drinking water, increasing Parkinson's risk for the surrounding area.[17]

Golf Course Pesticides, Drinking Water & Parkinson's Risk
Parkinson's Foundation; Aug 05, 2025

In a previous chapter I have presented facts and figures about the massive amounts of lead arsenate that were previously used on golf courses. I believe the previous use of lead arsenate is probably causing some, or most of, the increased risk of a Parkinson's Disease near golf courses. However, there were many other types of pesticides, fertilizers, and insecticides used in large amounts on golf courses, so it may not be possible to narrow it down to specific chemicals.

Many scientific studies have studied the statistical correlations between pesticides used on golf courses and various disease states. However, since there are only a few older records of the amount of lead arsenate used on each plot of land, the positive correlations between more modern pesticides and disease states may actually be due to prior lead arsenate use, which also correlates with more modern pesticide use.

Chapter 10 – The Great Divide: Red and Blue States

In the United States the political affiliations of states are often characterized as being red state or blue state, reflecting the views of the two main political parties, Republicans and Democrats. The red states are those states that tend to vote for conservative Republican candidates, and the blue states are those that vote for the liberal Democrats.

In recent years the Republican Party, led by Donald Trump, has greatly increased its efforts to demonize people from other countries. The views and actions of Donald Trump regarding people of other races and cultures is shockingly similar to Adolph Hitler and the Nazi's.

For example, the next figure from the Nazi era depicts Jews kidnapping and killing Christian children in order to drink their blood in a ritual murder. This sort of imagery was fairly common in that time, often appearing in mainstream newspapers and magazines, even though it obviously wasn't true. This sort of material served the Nazis well in their violent attempt to exterminate all the Jewish people from the face of the earth.

In an absurd claim bashing immigrants, at the 2024 presidential debate, Donald Trump said immigrants were eating cats and dogs. He also routinely calls illegal immigrants criminals and murderers who are invading the United States. The differences between Hitler's worldview and Trump's are mostly in their choice of ethnic minorities to blame for these imaginary problems.

It is so hard for me to believe that any rational person would hold such violent, vile, and immoral views. I have come to the conclusion that all those people are simply irrational and immoral. This caused me to wonder why all these people are irrational and immoral, and why they are generally located in rural areas.

Figure 18 Antisemitic depiction of Jews murdering children and drinking their blood.[18]

"**They're eating the dogs,** the people that came in, **they're eating the cats... They're eating the pets** of the people that live there, and this is what's happening in our country, and it's a shame."

Figure 19: Donald Trump bashing immigrants at a 2024 presidential debate.[19]

This led to my great eureka moment when I realized that geographical areas that tended to vote for conservative Republicans were also areas that were likely to have drinking water that is sourced from well-water. Then it all clicked together like a giant jig-saw puzzle. The areas where lead arsenate is contaminating the drinking water are the same areas that tend to vote for conservative Republicans!

I started to compare maps of voting patterns to maps of contaminated well-water areas and they were very similar. The areas of the United States that used large amounts of lead arsenate, and were sourcing drinking water from wells, were the same areas that were voting heavily Republican. The areas of heaviest lead arsenate use in the past were often areas where cotton, tobacco, and fruit trees were grown. Rice is sometimes grown in these areas today. These areas include the southern states, Florida, the Mississippi River watershed, and the Central Valley of California.

The billions of pounds of poisonous lead arsenate pesticide that was applied worldwide in the last century is now seeping into ground water and currently poisoning vast areas of the world. This is causing a large percentage of the world's people to suffer from the effects lead poisoning, and they will often exhibit symptoms of mental retardation, xenophobia and a propensity to violence. These poisoned people are highly paranoid and feel that people from other groups are violent criminals out to get them, and maybe even eat their pets or children.

All over the world These poisoned "leadhead" victims often support politicians who share their paranoid worldview. Those politicians are often dictators, or want to be dictators, so they can "defend" their people from their paranoid delusions that other groups of people are out to get them. For example, Donald Trump made this ridiculous claim about transgender people:

> *"No transgender, no operations —
> you know, they take your kid —
> there are some places, your boy
> leaves for school, comes back a*

girl. Okay? Without parental consent." He added, *"At first, when I was told that was actually happening, I said, you know, it's an exaggeration. No: it happens. It happens. There are areas where it happens."* [20]

The idea that these are rational thoughts is completely absurd. Clearly anyone who believes these crazy ideas is not a rational person and they have no business running anything of importance.

Imagine that you are in a small group of administrators in charge of hiring people to work at a large business. In one job interview the applicant rants about these involuntary sex change operations at schools, and how dirty, filthy, criminal immigrants were eating their neighbor's cats and dogs. How long do you think it would take before you politely asked him to leave, and then throw his job application in the trash? If another member of your group thought the applicant would be great for the job, how long would it take before he or she was removed from their job?

Yet this is exactly what Donald Trump was doing at the time. He was running for president and these presidential debates and campaign speeches are essentially job interviews with the people in charge of hiring him (by voting for him).

Chapter 11 – Looking Back and Going Forward

It is often said that hindsight is 20/20. The things we know today make it clear that the massive use of lead arsenate as a pesticide was the greatest disaster to affect mankind in the entire history of the earth. By directly causing both World Wars and the holocaust, and many other violent deaths, this poison killed more people than any other man-made or natural disaster, perhaps even more than all the other disasters combined.

The earliest use of lead arsenate pesticide may have been due to a desire to increase food production. In that era, starvation, malnutrition, and diseases were common problems. It may have seemed reasonable to use this poisonous pesticide to increase production of apples, oranges, and other fruits. In that era, most farmers and scientists thought that all of the pesticide could be washed off the fruits with a slightly acidic water bath.

In the USA, the USDA (US Department of Agriculture) published many glowing reports about how this new wonderful pesticide was increasing crop production and making populations healthier, and farmers wealthier, by increasing the amount of fruits available to the rapidly growing urban populations. Before lead arsenate came around most apples and other fruits would rot before they made to to the urban consumers.

[Note: Most of my analysis about *why* this lead arsenate poisoning disaster happened, and why it wasn't stopped, is based mainly on US data and documents. Other countries probably had similar circumstances and reasons why they didn't ban lead arsenate, but this was not substantially investigated by me.]

Ignorance of the toxic effects of lead arsenate in the early 1900's was probably to blame its early use on farms. However, the USDA continued to promote the use of lead arsenate, even after it became clear that it was very toxic, by continuing to assert that it could be easily washed off apples and other fruits with a simple water bath. In publication after publication, the USDA wrote extensively about the health effects of lead arsenate on fruit trees but I didn't see a single word about the pesticide's toxic effects on the health of humans. For example, this quote from the USDA was typical for this era.

Lead arsenate is the only dependable stomach insecticide for the control of codling moth and various other chewing insects that attack apple, pear, peach, and other fruit trees. In some parts of the country, notably Colorado and Washington, the codling moth is less susceptible to control by lead arsenate than in other parts of the country. This may be due to a resistance on the part of the insect that has been developed by years of arsenical spraying. Owing to the great difficulty in controlling this insect it has become necessary to apply 8 or 10 sprays of lead arsenate in order to produce marketable apples. This heavy spraying results in the accumulation upon the fruit of quantities of arsenic exceeding 0.01 grain arsenic trioxide per pound, the maximum quantity permitted under the regulations of the British Government, and consequently the fruit must be washed or wiped in order that it may be marketed.[21]

--From the *USDA Yearbook of Agriculture 1928*

[Note that since 1 grain is equal to 65 mg, then 0.01 grain/pound = 1.43 mg/kg. This is over a thousand times greater than current WHO (World Health Organization) provisional guideline values for lead in water.]

This maximum quantity of arsenic permitted by the British government (above) still results in a lethal dose of arsenic (about 140 mg) in every 100 kg (220 lbs) of apples. Since arsenic, like most heavy metals, accumulates in the human body over time, a lethal dose of arsenic could be absorbed by eating just two apples a day for a year.

A common phrase at that time was "an apple a day keeps the doctor away", but a more accurate one would be "an apple a day keeps the doctors well paid".

By the early 1930's the toxic effects of lead arsenate on humans was beginning to be realized, but these health concerns were ignored by those in the government in order to protect the fruit growing industry.

In one experiment in 1931 two researchers at the USDA (E.M. Nelson and H.H. Mottern) tested the effects of orange juice from trees that were sprayed with lead arsenate on guinea pigs. They reported that:

> *Twelve animals were fed the normal orange juice, and 12 the juice from sprayed trees. Two of the last died within a few days and are not considered in analyzing the results.*[22]

As a scientist, it seems extremely absurd that anyone would ignore the fact that one sixth of your test animals died after drinking orange juice from trees sprayed with lead arsenate. If those two researchers had acted with a little bit of integrity after finding out that lead arsenate pesticide was poisoning nearly everyone in the world, they should have reported this fact and they probably would have prevented the first wave of "Deranged Nazi Syndrome" and the 100 million deaths caused by WWII, The Holocaust, and other wars of the mid-1900's. This would have also prevented the current second wave of "Deranged Nazi Syndrome" mainly due to waterborne lead arsenate.

One of the important methods used in the successful campaign to eradicate the Mediterranean fruit fly from Florida was the application of a poison spray on which the adult flies fed. The most effective bait spray available in this emergency contained lead arsenate as the poison. Although it had been known for many years that arsenic hastened the maturity of citrus fruit by reducing the acid content and that its use had, in some instances, rendered the fruit unpalatable, t**he menace of this insect pest to the citrus industry of Florida and to fruit production elsewhere in the United States was so great that the authorization for the use of this poison was fully justified, and its use was approved by responsible State officials.**[23]

These scientists and governmental officials were probably under political pressure by agricultural groups to support the lie that using lead arsenate pesticide was a safe and effective way to increase crop yields and decrease starvation and malnutrition.

Now that I have looked back at the ghosts of Christmas-past, it's time to look at the ghosts of Christmas-present and the ghosts of Christmas-future, and see how we can change our ways.

The insane mass shootings and terrorist attacks that are currently affecting many countries, and the Nazi style political xenophobia and paranoid delusions that are infecting many countries today, are simply the result of lead arsenate poisoning slowly creeping into well-water supplies, and causing this second wave of "Deranged Nazi Syndrome".

It may be hard for some people to accept the enormity of the disasters that were brought about because of lead arsenate use, and the 100 million people that were probably killed because of it. Denial of the facts and attacking the messenger may be an emotional escape for some people, however we need to clearly see the past so we can fix the future.

Even though the past use of lead arsenate pesticide had such dreadful and consequential effects upon the world, there is a brighter future in store for all of us as long as we learn from the past and take corrective action now.

A future without any wars or violent crime may seem like a pipe dream, but we can certainly get most of the way there by simply eliminating our exposure to lead from food, leaded gasoline, and well-water.   The fact that violent crimes has decreased by half in cities worldwide as leaded gasoline was eliminated shows us a clear path to world peace and an end to war.

Postscript

The first edition of this book was rushed to completion by Christmas Day 2025. It is my Christmas gift to the world. Please share this book with your friends and family.

Next year I hope to write a more complete edition of this book that includes more technical information about removing lead and arsenic from our farmlands, food supplies, and drinking water. I also want to expand on many chapters in this book, especially those concerning the correlation between lead exposure and political views and voting patterns. However, I felt that the information contained in this book is of so much importance that I could not wait any longer before publishing it.

Every Christmas, people pray for world peace and prosperity, but this rarely happens. In this book I have shown the world the bad news about past agricultural practices, and explained how this caused the world to explode in an orgy of murder and violence that still affects us today.

It is very obvious to me that a rational path to world peace and prosperity needs to start with removing lead and arsenic from our food and drinking water supplies.

If we take action today there is a good chance that when next Christmas comes around people will not be praying in vain for peace and prosperity, but will be giving thanks for all the peace and prosperity that has swept over the world.

--- Wishing You a Merry Christmas and Peace and Prosperity For All,

Chris Baba, December 25, 2025,
Frederick Maryland USA

Chris Baba 2009

About the Author

Chris Baba has a Master's degree in Molecular Biology / Biotechnology from Johns Hopkins University and a Bachelor's degree in Biochemistry and Molecular Biology from the University of Maryland-Baltimore County.

He has spent much of his life working in laboratories doing medical research. He has worked at the University of Maryland School of Medicine and Johns Hopkins University School of Medicine. He also worked for the Howard Hughes Medical Institute, the NCI-Frederick Cancer Research Facility, and MedImmune.

Chris Baba had to stop working when he was stricken with a genetic disease, Loeys-Dietz Syndrome, which required open heart surgery. Now he spends his time at home pondering the mysteries of the universe and writing open source computer programs in C++ and Qt. He is amazed by AI and dabbles in it with his home computer.

Chris and his Wife Megan live in the mountainous town of Frederick Maryland, where he did a lot of mountain biking when he was younger. Now he uses his electric bike to exercise his two dogs.

How to get this book in different formats.

This book is copyrighted by me but licensed under a permissive license that allows copying or sharing electronically as long as you don't do it for a profit.

I have posted this book in various electronic formats (PDFs and Ebooks) that are free for you to download at my GitHub Account. Please tell your friends to visit this site for free downloads of this book:

https://github.com/Chris-Baba/Bad-Apples-Book

Bad Apples, Bad Water is licensed under Creative Commons Attribution-NonCommercial-NoDerivatives 4.0

International.license To view a copy of this license, visit https://creativecommons.org/licenses/by-nc-nd/4.0/

To get Kindle or printed versions of the book at a minimal cost (hardcover and paperback) visit Amazon.com. Please remember that I can sell these books for a profit, but you can't.

References and Notes

[1] *Aircraft Year Book 1925* - Aeronautical Chamber of Commerce of America 1925
https://www.aia-aerospace.org/wp-content/uploads/the-1925-aircraft-year-book.pdf

[2] *USDA Agriculture Statistics 1941*
https://babel.hathitrust.org/cgi/pt?id=nyp.33433108713425&seq=8

[3] *The History of Lead Arsenate Use in Apple Production: Comparison of its Impact in Virginia with Other States.* Therese Schooley et.al.,Journal of Pesticide Safety Education Volume 10, 2008 https://www.aapse.org/wp-content/uploads/1-195-1-PB.pdf

[4] *Arsenic.* Collaborative for Health & Environment
https://www.healthandenvironment.org/environmental-health/environmental-risks/chemical-environment-overview/arsenic
Retrieved 2-16-2025.

[5] *USDA Yearbook of Agriculture 1930*, Milton S. Eisenhower, pg 348 https://archive.org/details/yoa1930/page/n1/mode/2up

[6] US Geological Survey (USGS)
http://USGS.gov/media/images/pesticides-groundwater-can-eventually-contaminate-well-water

[7] *Lead and Lead Poisoning in Antiquity* Jerome O. Nriagu 1983 (page 6) Environment Canada, National Water Research Institute Burlington, Ontario Copyright© 1983 by John Wiley & Sons Inc.,
https://dl.tufts.edu/concern/pdfs/6395wk23b

[8] *WHO Guidelines for drinking-water quality: fourth edition incorporating the first and second addenda* (21 March 2022)
https://www.who.int/publications/i/item/9789240045064

[9] Although the total amount of lead emitted from leaded gasoline is larger than the amount applied as lead arsenate pesticide, the chemical nature (metallic vs. salt) of each type of exposure affects the bioavailability.

[10] *Environmental causes of violence* by D.O. Carpenter and R. Nevin *Physiology & Behavior* 99 (2010) 260–268
https://www.sciencedirect.com/science/article/abs/pii/S0031938
40900300X?via%3Dihub

[11] *The Bell Curve: Intelligence and Class Structure in American Life* by Richard J. Herrnstein and Charles Murray (Free Press, 1994)
https://en.wikipedia.org/wiki/The_Bell_Curve

[12] *Lead toxicity: a review* Ab Latif Wani, Anjum Ara Jawed Ahmad Usmani Interdiscip Toxicol. 2015; Vol. 8(2): 55–64.
https://pmc.ncbi.nlm.nih.gov/articles/PMC4961898/

[13] *The end of the Flynn effect?: A study of secular trends in mean intelligence test scores of Norwegian conscripts during half a century.* By Sundet, J.; Barlaug, D.; Torjussen, T. (2004). Intelligence. 32 (4): 349–62.
https://www.sciencedirect.com/science/article/abs/pii/S01602896
04000522?via%3Dihub Graph copied from Wikipedia
https://en.wikipedia.org/wiki/Flynn_effect

[14] *The great American IQ panic of 1917* Community Living Magazine 21 April 2018 — Spring 2018
https://communitylivingmagazine.com/great-american-iq-panic-1917/

[15] Hitler's Micrographia Original Source: Lieberman, A. *Adolph Hitler had Post-encephalitic Parkinsonism.* Parkinsonism & Releated Disorders 1992;2:95-103. Figure shown was copied from: *Adolf Hitler and His Parkinsonism* Kalyan B. Bhattacharyya *Annals of Indian Academy of Neurology* 2015;18:387-390
PubMed Link

[16] *Adolf Hitler and His Parkinsonism* by Kalyan B. Bhattacharyya; *Annals of Indian Academy of Neurology* 2015;18:387-390
https://pmc.ncbi.nlm.nih.gov/articles/PMC4683874/

[17] *Golf Course Pesticides, Drinking Water & Parkinson's Risk*
https://www.parkinson.org/blog/science-news/golf-courses
Parkinson's Foundation; Aug 05, 2025

[18] *Antisemitic Imagery and Caricatures* (May 2020), Antisemitism Policy Trust https://antisemitism.org.uk/wp-content/uploads/2020/07/Antisemitic-imagery-May-2020.pdf

[19] https://www.nbcnews.com/politics/2024-election/trump-pushes-baseless-claim-immigrants-eating-pets-rcna170537

[20] https://www.cnn.com/2024/10/26/politics/fact-check-trump-rogan-children-gender-affirming-surgeries/index.html

[21] *The USDA Yearbook of Agriculture 1928* https://archive.org/details/yoa1928/page/n1/mode/2up

[22] *Effect of Lead Arsenate Spray on the Composition and Vitamin Content of Oranges* American Journal of Public Health; E. M. Nelson and H. H. Mottern Bureau of Chemistry and Soils, U. S. Department of Agriculture, Washington, D. C. *Read before the Food, Drugs and Nutrition Section of the American Public Health Association at the sixtieth Annual Meeting at Montreal, Canada, September 15 1931. https://ajph.aphapublications.org/doi/epdf/10.2105/AJPH.22.6.587

[23] *Effect of Lead Arsenate Insecticides on Orange Trees in Florida*, United States Department of Agriculture - Technical Bulletin 350; R. Miller, I. Bassett, and W. Youthers; February 1933 https://ageconsearch.umn.edu/record/163537/files/tb350.pdf

www.ingramcontent.com/pod-product-compliance
Lightning Source LLC
Chambersburg PA
CBHW061747020426
42331CB00006B/1384